MAINTENANCE ON THE MP40, PPSH41, M3A1, AND STEN MKII SUBMACHINE GUNS

Glenn J Fleming Jr.

Again, thank you to Pat, Jamie, and many others who have helped me in this endeavor. My grammar is horrible on the best of days and they try consistently to reign it in so it's readable. You don't know how many curse words they edit out.

I'd be remiss if I didn't mention Butch, at the tender age of 14 he placed an MP40 in my hands to shoot and I've been hooked on machine guns since. Thanks man.

Also being my second book, I'd like to dedicate it to all those who bought my first book. I can't thank you enough! You are all my inspiration to continue this new little "hobby" I have. Well that and money.

I hope this, and future books, will give you the knowledge you seek.

Glenn J Fleming Jr. M.O.D.

Thoughts on stuff

Welcome to the wonderful world of blow back operation! Simply put, for every action there is an equal and opposite re-action. It really is that simple. There are many designs that use this principle, but for the purpose of this book I've selected four that I think will give an overall glimpse into their function and methods of repair.

My first book was designed to be a general "guide" into the thinking behind working on this stuff. These next few books will cover specific guns. Don't get your hopes up though, as there won't be much in the way of history on the guns. There are already countless books that deal with that. I'm going to be talking about the details of operation and repair. I like wise won't be giving much airtime to "this part is broken, take it out and replace it" that, afterall, is common-sense stuff.

Lastly, I've gotten emails wondering about the spacing so I may as well address it here. These books are envisioned to be used while you work on a gun. I find it much easier to read along when there is a bit of space.

With all that having been said.... Enjoy!

1 THE MP40 SUBMACHINE GUN

"SCHMEISSIER"

The MP40 is the first machine gun I ever shot and still remember it fondly. That's no small feat for me as I have the memory of a gold fish and don't even remember what I had for breakfast this morning.

Often erroneously called the "Schmeisser", this gun was developed in the late 30's, but not by Hugo Schmeisser.

The idea behind the gun was to simplify the manufacturing process of the MP38. Where the MP38 used milled parts the MP40 used stamped metal, thus saving time and money.

The gun is simplicity itself. The bolt rides in the tubular receiver stripping rounds from a magazine. The fire control group (FCG) only has three parts with a couple pins and one spring.

So, if its so easy why write a book on it?

Well, there are things you will need to be aware of when working on one and there is some stuff that should be addressed about firing it as well. It's my job (Is this really a job?) to make sure those things get across.

So, let's get to it.

Let me first address that the nomenclature for the parts will be from John Baums translated manuals; after all they are directly from the German publications and that seems a pretty good source: right? They can be found at www.germanmanuals.com.

Next, before you do anything to an MP40, be VERY careful with the Bakelite furniture. That stuff is very old and if you break it, replacing it can be difficult to say the least.

All that having been said, let's take one of these apart!

MP40 Basic Disassembly

1. Ensure gun is unloaded and bolt is in the forward position

2. Pull down and rotate the breech pin 90 degrees. This will loosen the upper receiver from the lower.

Above you can see the breech pin.

3. With the trigger held down rotate the lower receiver on the upper counterclockwise, when viewed from the rear until pistol grip is in the four o'clock position.

4. Pull lower to the rear and off the receiver.

5. Pull bolt group to rear and out of gun being careful not to drop telescoping spring housing.

6. Remove telescoping spring housing from bolt head.

That is the basic field stripping of the gun. Don't you wish all guns could be that easy? I have a sneaking suspicion that most of you already knew that though so let's get a little more in-depth with taking this sucker apart.

MP40 Expanded Disassembly

Before we start let me give a word of warning: The MP40 receiver tube is susceptible to denting and deformation. When you are removing pins make sure you support the weapon correctly. Never put the receiver tube in a vice without some sort of interior support in the receiver to keep it from deforming!

Upper Receiver Section

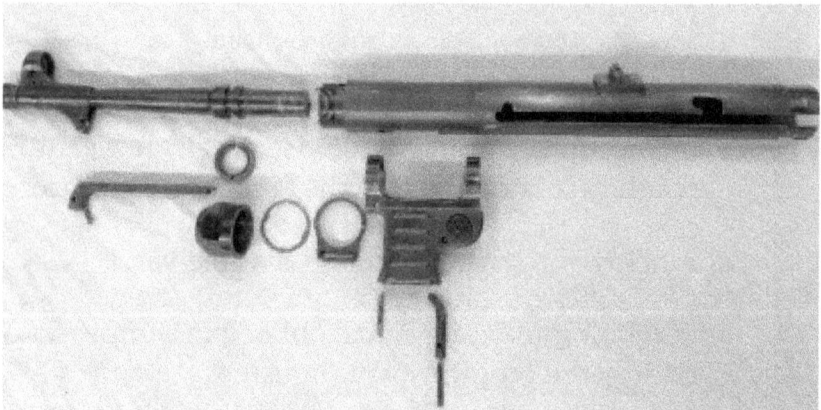

The parts of the upper receiver laid out, complete with pins.

To begin disassembly of the upper receiver, first put the BARREL in a vice with soft jaws. I prefer the ones made of a hard polyurethane for this. They have a half round notch cut on both sides running the length of the jaws. That holds the gun securely and makes it much easier to work on.

Once that is done, look to the area between the barrel nut and the collar (the front sling attachment part). There is a thin shim placed there. This shim is called the safety ring. Notice there are notches in the collar and the barrel nut cut out, top and bottom of the receiver. The safety ring should be bent in those areas to prevent the barrel nut from rotating. These areas on the safety ring will have to be bent back to be able to turn the barrel nut. This ring can be reused to a point, but they do break through work hardening if taken on and off to many times. Unfortunately, I don't know of anyone making new rings so if one breaks its time to make one.

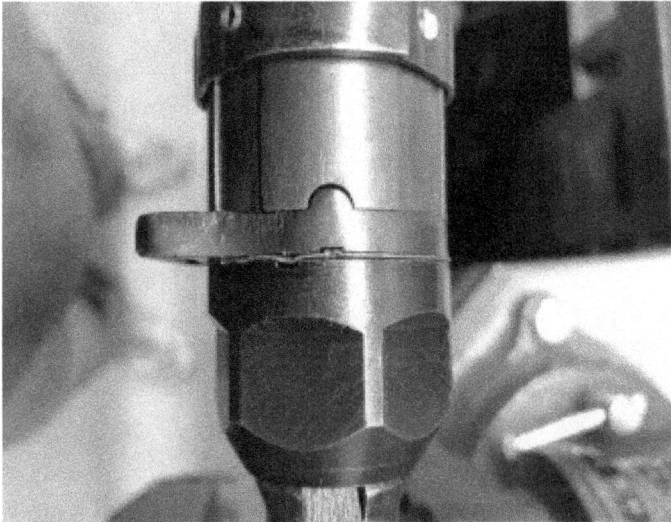

Seen here just above the barrel nut is the safety ring. Notice how the rings goes into the gaps cut into the barrel nut and the collar.

Once that is done place masking tape on the flat surfaces of the barrel nut. This will help prevent scratching the finish. At this point take your favorite big ass crescent wrench and commence to loosening it up. If there is a lot of resistance, stop and soak the barrel nut in automatic transmission fluid mixed with diesel at about a 50-50 mixture. Let it soak for a day or so and that should help with getting the nut off. Remember, that nut may have been on there for seventy plus years.

When you take the nut off there is a split ring underneath it. This will fall on the floor and you are guaranteed to lose one half of it if you are not ready; so put your hand under that area to catch it.

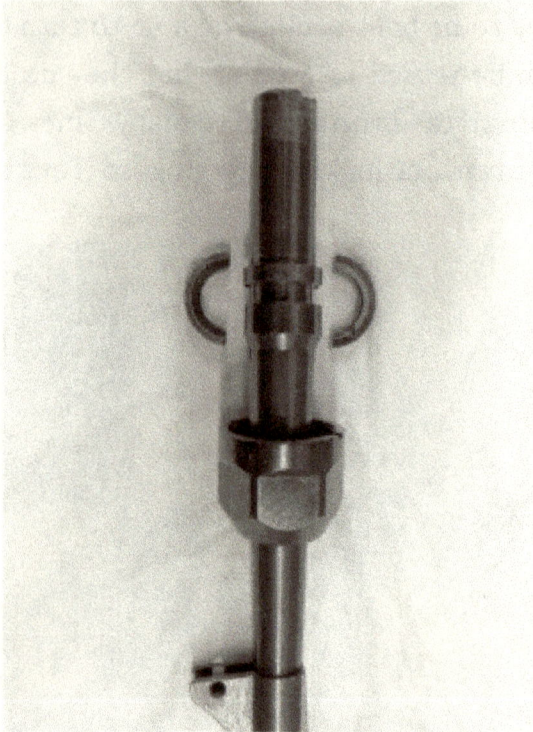

The split ring placed around its location on the barrel. Barrel is pointing down.

Next, remove the collar. It's a straight pull off type part but this can sometimes be a pain as they are often tight. Just take your time and it should come off. Be careful not to screw up the trunnion threads when doing this.

Now it's time to start on the magwell. Again, be very careful not to dent or deform the receiver tube.

There are two pins that need to be removed. One of the pins holds in the ejector, the other is at the front of the magwell. Take off both to remove the magwell.

It doesn't really matter which pin is removed first; but just bear in mind that the pins are flared on the ends. They should be reused so try not to deform the flared ends too much.

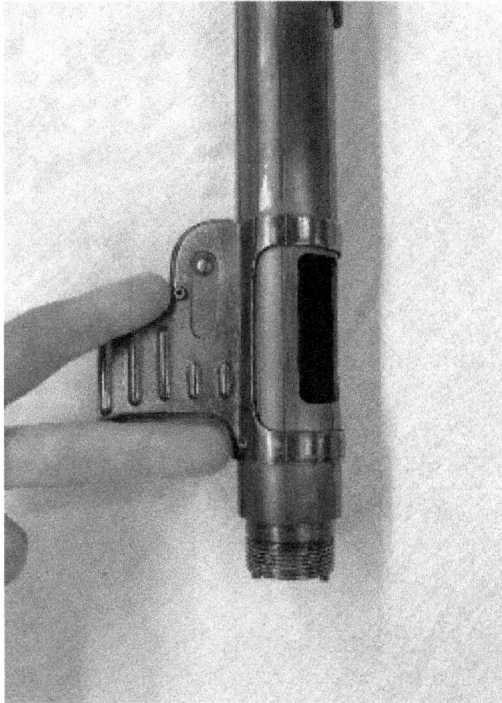

My high-tech pointing devices show the location of the two pins in the magwell. The rear pin also holds in the ejector

Regarding the ejector pin; once the pin is removed you will have to rotate the ejector 90 degrees clockwise when viewed from above. It won't clear the ejection port if you don't. Once rotated; pull up and out the receiver.

In this picture the ejector retaining pin has been removed and the ejector has been rotated so it can be removed.

After both pins and the ejector are removed, just slide the magwell off the front of the receiver.

The magcatch is riveted in the gun and normally there isn't a need to replace it. If it does need to be replaced,

Drill the rivet, install the new part, and re-rivet. There aren't any surprises in this area.

That is as far as you can disassemble the upper receiver unless you need to change the trunnion. Being as the trunnion is permanently affixed to the receiver tube removal is a challenge. I've never seen an original made trunion removed where the receiver has been saved.

I've only removed one and that was from a post sample gun. Luckily the welds were easy to see as they turned a different color in the bluing process. That made it relatively simple to drill out the spot welds and remove the trunnion.

Lower Receiver Section

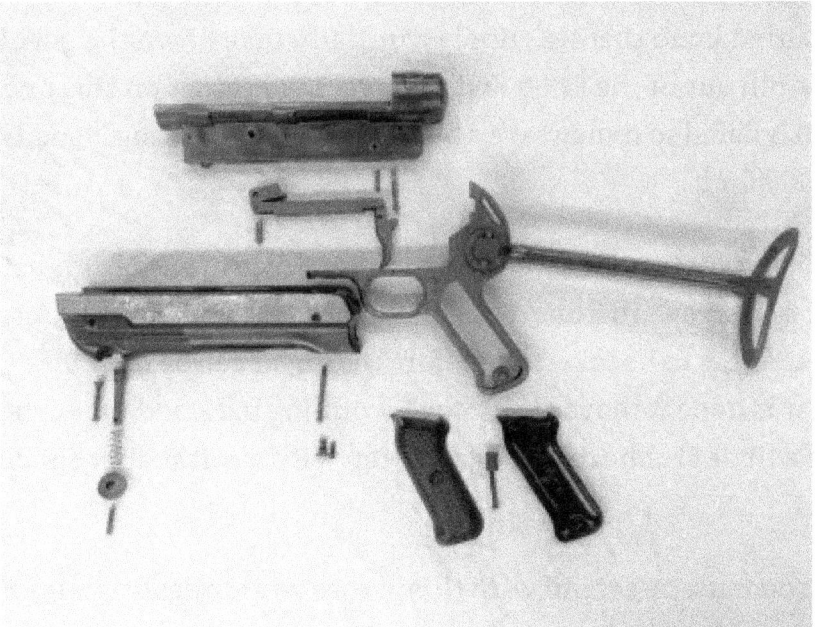

The parts of the lower receiver laid out, complete with pins.

Again, a word of warning before we begin, be very careful with the Bakelite. It can be very fragile and is both hard to replace and expensive.

The first thing to do when disassembling the lower is remove the pistol grip halves. They are held in place by a screw that runs through them and a knurled knob on the other grip that acts as a nut. Sometimes the nut will move in its hole. I've found that most times just putting some finger pressure is enough to make it stop turning so you can take out the screw. If it continues to turn, another option is to press on it with needle nose pliers. In either case once the screw is removed you can remove the grip halves. There should be a spacer placed between them so watch out for that.

Next you will want to remove the breech pin. This is the knurled knob that was used to initially take down the gun. Push the pin out of the knob and remove. Two things on this one; the pin is flared so it may take a bit of force and it's spring loaded so be ready for it.

Now we turn our attention to the pistol grip frame. It is held on by two screws in front. The smaller of those screws is a lock screw that needs to be taken out before the larger can be removed. Once that is done remove the frame by pulling back and down. It may be a little stubborn but a gentle tap with a soft mallet should do it.

You must be careful with this part as you are dealing with Bakelite.

Take out the two screws holding the large Bakelite shell on and remove it by pulling down and forward. You may have to spread it a little bit, if so do it gently.

Taking the FCG out is as easy as removing the two pins that hold it in. There is a spring with a domed pin that is on the trigger that will come loose when you remove the pins. Make sure you watch out for that.

The buttstock is held in by a single pin with a pretty stout spring underneath. Once the pin, push button, and spring are removed spread the "arms" of the stock and remove.

That is disassembly of the lower receiver.

Bolt Section

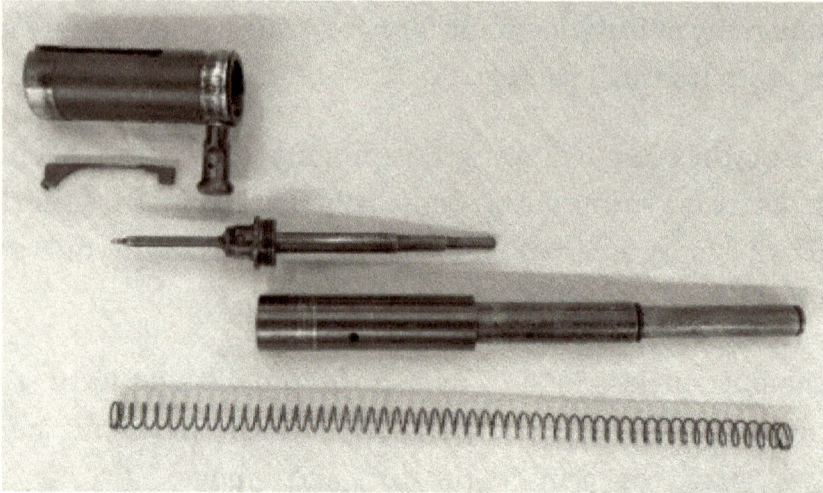

The parts of the bolt group disassembled. The buffer section housed in the area behind the firing pin is still assembled here.

The bolt and telescoping spring housing are essentially held together by the pressure of the spring. The telescoping spring housing assembly sits in the bolt head and the lower receiver pushes the rear of it to maintain constant compression.

To take the bolt head off, just pull and set aside. In the bolt head is the extractor. Directly across from the extractor in the bolt head is a hole. Put a punch in the hole and you can push out the extractor.

To take apart the telescoping spring housing simply unscrew the firing pin assembly. I shouldn't have to say this, but I know someone will lose and eye and sue me so:

Warning: The telescoping spring assembly has a spring in it and is under spring pressure.

Finally, we strip the buffer assembly down. Simply remove the pin holding the firing pin in. Pull the firing pin off to expose yet another spring and the plunger.

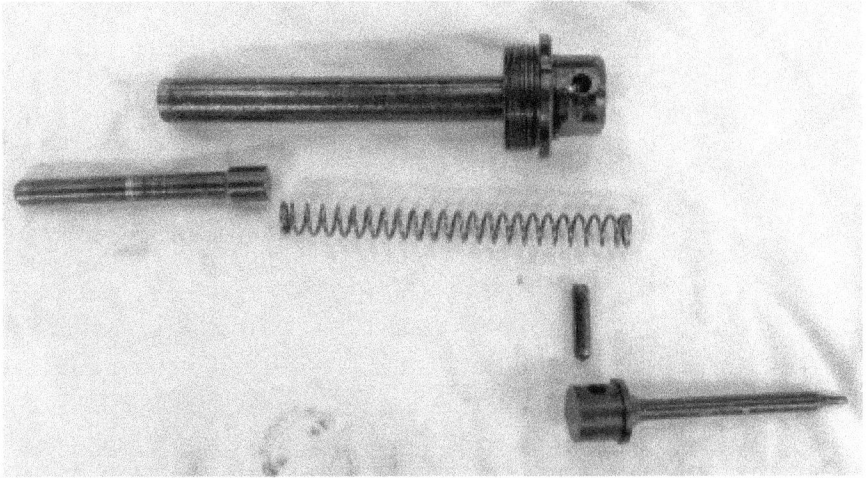

Here are the parts of the buffer assembly broken down.

There you have it. The complete gun taken down to its components. I really can't believe I filled as many pages on this as I have. It's got to be a record!

Now we turn our keen gunsmith eyes to malfunctions and repairs. I should mention some of these were covered in my first book, but I don't know if you have bought my first book, now do I?

MP40 Malfunctions and repairs.

Starting with ejection problems, there can be a few issues: ejector, the extractor, chamber, or even a damaged receiver tube.

If you are getting stove pipes, or the rounds are not ejecting from the bolt face at all, I'm willing to bet your ejector "ist kaput" as the Germans would say. The easiest fix is replacement, especially if missing a major part of the ejector. If, however, it's only a chip in the flat of the ejector you may be able to file the face flat and restore it to its former glory. Since you are doing that, you could put a very slight edge to it to aid in it's ejection direction.

When firing if your case suddenly stays in the chamber even though the bolt is back, that's usually caused by either a bad extractor or a bad chamber.

If the extractor is suspected, first check to be sure it isn't just carbon build up or another type of fouling impeding the extractor from doing its job. If its not a matter of dirt and it is indeed broke, time to find a new one.

The chamber could have a burr or rust in it. This would slow recoil and hurt extraction and ejection. A quick fix for this is to use a chamber hone of the right size, judiciously doused in oil, to smooth the chamber out. Don't go overboard though, as you don't want to suddenly find out you have a 45 caliber chamber after honing.

Needless to say, if there are no obvious signs of damage in any of the above cases look at your recoil spring. It could be going bad or you could have some debris in your telescoping tube.

If a damaged receiver tube is the case, there are a couple things that can be done.

The receiver tube on the MP40 isn't very thick and is susceptible to denting. This, of course, really gums up the works. If the damage on the interior is slight, you can use a lapping compound to smooth out the area. A file or sandpaper wrapped around a wooden dowel would work if you need to be slightly more aggressive.

Bear in mind though the tube is already thin enough. You don't want it to be much thinner.

If the dent is larger than you are comfortable repairing in those methods, you will need to build an adjustable backer to aid in the repair.

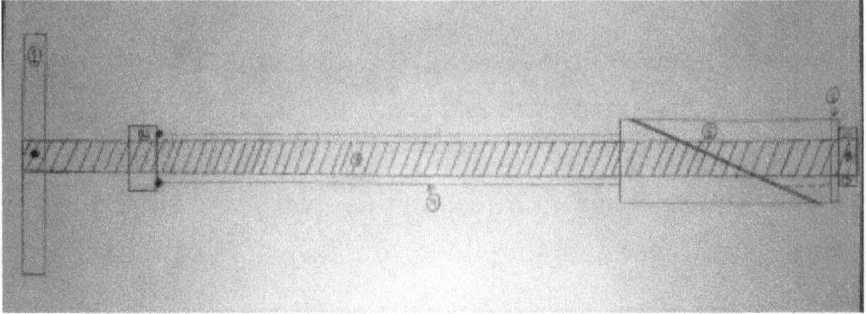

The backer I've made to remove dents from tubes, including an MP40 receiver tube. From left to right the parts are.

1. Is a metal rod pinned or welded to the threaded rod (3). This makes a "T" handle that will be used to turn the threaded rod.
2. A nut welded to the tube (4) surrounding the threaded rod (3).
3. Threaded rod
4. Tube surrounding the threaded rod.
5. The two backer halves. They are made of round stock that has been milled at an angle so they can slide up when drawn together by means of turning the threaded rod (3). The half on the end (RH) should have its middle hole slightly elongated to the bottom so it can slide up.
6. Washer
7. End cap welded or pinned to the threaded rod (3).

The design of the rod is simple enough. You basically want something that can be tightened up against the dent as you work it out. The trick is how you work the dent from the outside. Once the backer is in place, with a brass hammer, tap around the edge of the dent "pushing" it towards the center of the dent.

When I say "tap" I don't mean use a 1-ounce hammer and work it with two fingers, and I certainly don't mean go after it like Hephaestus having a bad day.

Just use enough force to work the metal but not make the issue worse.

Once you can expand the backer again retighten it against the dent and repeat the hammering. Keep doing this until the dent is gone.

When the dent has been removed there will likely be some deformity of the metal where it was. This can be addressed with a file or fine sandpaper depending on the need.

Most feeding issues I've seen have been due to bad magazines, usually the feed lips are dented. It may be possible to bend the feed lip back into place, but it is often an exercise in futility. I've always found it easier to buy a new magazine even though they are a bit on the expensive side.

I have seen recoil springs, firing pins, as well as extractors break, but that is basically the only other malfunctions I've seen. In the case of parts breakage just buy a new part. Worst case scenario you will have to make one.

MP40 Reassembly

I'm just going to touch on this for a brief minute. There really aren't any special instructions to it, but there are a couple tips I can give. First when re installing a new firing pin, it recesses in the head of the buffer housing, it sometimes helps to put the firing pin in a freezer overnight. Most are very tight and doing this allows it to shrink a bit and make the job a tad easier.

Finally, though assembly is pretty much the reverse of disassembly, ensure that when you place the lower receiver on the upper receiver that you center the telescoping rod in the cup of the lower. It should automatically sit there but I have seen one that didn't so it's better to be sure. After all at the time of this writing, these guns are going for around $18,000.00! With that kind of money, you don't want to mess anything up.

Firing the MP40

Now the fun part, shooting the gun!

When I take someone to shoot machine guns for their first time, I usually take the MP40. I like to call it my starter machine gun. It's so forgiving when shooting practically anyone can handle it correctly right off the bat.

With a slow rate of fire, firing a 9mm round, and the gun having a little weight to it, it's just a pleasant gun to shoot.

There are only a couple "gotchas" when shooting the MP40, things that you should be aware of when firing it for the first time.

The first of which is don't hold the magazine while firing. While it doesn't always produce a jam, it sure has the possibility of doing so. The magazine in the MP40 does wiggle a little bit, so if you hold the mag while firing you can change the angle of the magazine, thereby changing the angle of the feed in the gun. If your gun has enough wiggle to it guess what?

Yep, that's right, a jam.

The next thing is the safety lock on the charging handle; placed there because the Germans aren't happy unless each part in a gun is comprised of at least three parts. The lock is the small piece of metal surrounding the charging handle that goes in and out, presumebly to lock the handle in the forward position. If accidently engaged, you won't be able to pull the bolt back without a decent amount of effort. If you do get the bolt back, it will drag while firing and could cause a light strike. Lastly, if the lock is loose on the bolt you could run into the problem of it engaging itself while firing, no bueno.

Thus endeth the MP40 section.

2 THE PPSH 41
"PAPASHA"

One of the most exciting guns you can work on and one that will always get attention at any shoot is the PPSH 41, a Russian made bullet hose. Capable of firing 1000 rounds a minute, it was exceedingly popular. Indeed, the Germans liked it so much, whenever possible, they would take them for their own use. There was even an adapter made to convert the gun to 9mm using MP40 magazines!

The gun is different from most other WWII blow back operation sub guns in that it has a provision for select fire. If you don't feel like wasting all your ammo in a very short time, just put it in semi and prolong the experience!

Of course, to the person repairing it that just means more parts, but never fear. Its actually very easy to diagnose and repair.

PPSH 41 Basic Disassembly

1. Ensure the gun is unloaded and the bolt is in the forward position.

2. Push forward on the rear catch while tilting up on the upper receiver.

3. Grasp the bolt and pull slightly back and up to remove the bolt, recoil spring, and buffer.

That's it. For general cleaning, that's as far as the gun needs to be broken down. Its ridiculously simple. We, however, don't like simple so we are going to confuse the issue and take it all the way apart!

PPSH 41 Expanded Disassembly

To break down the gun into its components start by removing the pin that holds the upper and lower receivers together. It's a two-part split pin design. Drive the split end of the pin out of the larger pin and after that has been done drive the larger pin out from the other side. Separate the upper from lower.

Upper Receiver Section

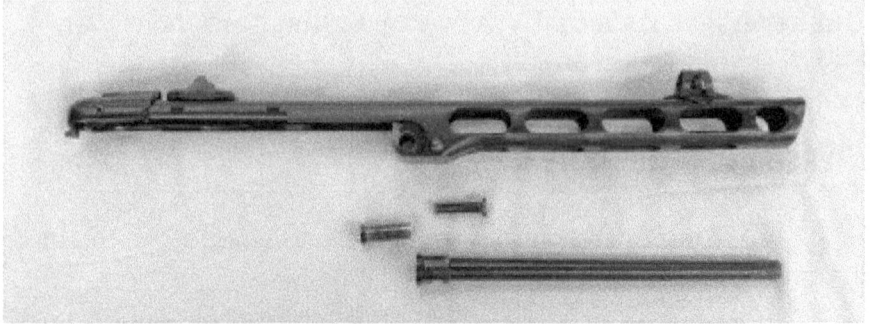

Upper receiver section of the PPSH with barrel removed.

When disassembling the PPSH, one can remove the rear take down latch to start, but it's not advisable. The pin that holds it on is flared and unless something is broken in the rear of the gun, it should be left in place. If the pin is removed look into replacement if possible. If not, ensure the existing pin is well flared at both ends so it holds securely. The last thing you want is that rear catch coming loose.

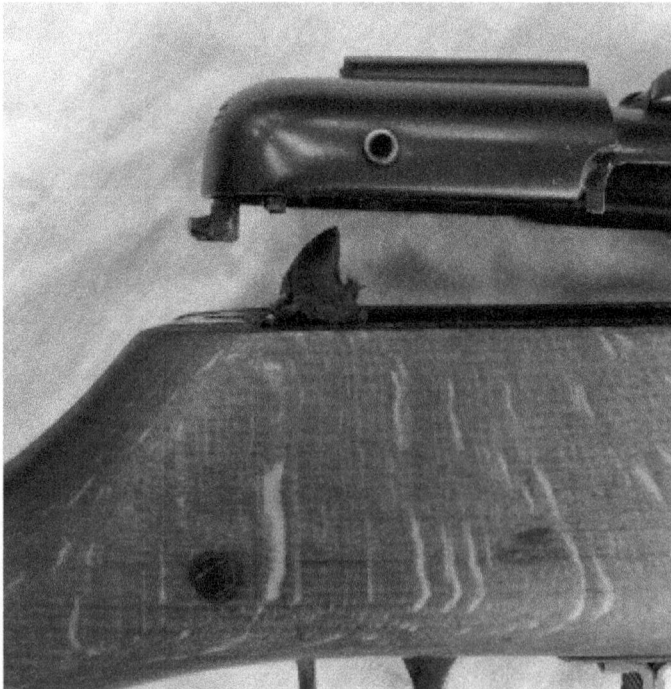

The rear take down latch showing the flared pin as well

Moving to the front of the receiver, take a brass punch and place it against the face of the barrel. Tap the punch with a hammer to drive the barrel out of the jacket and trunnion.

Again, simplicity itself.

Lower Receiver

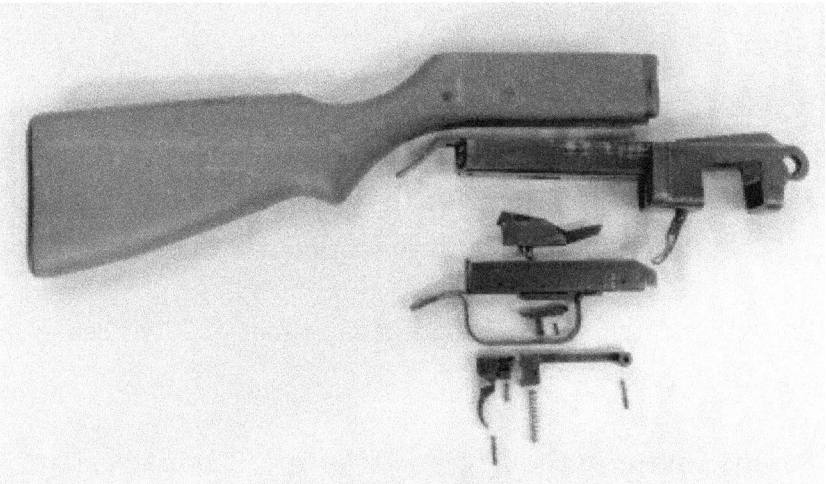

Lover receiver section of the PPHS exploded view.
Note mag release is still in place.

Now here is where we get to the fun parts!

Remove the rear screw from the top on the lower receiver just behind where the upper receiver latches on. This will split the top section from the bottom and free the buttstock. Simply drop down the bottom section to remove it. At that point, tilt the buttstock down and rearward.

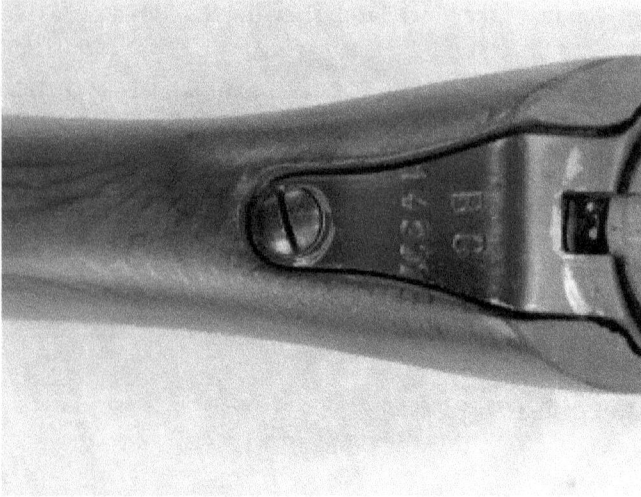

The rear take down screw. Located just behind the rear latch

Once the upper and lower parts of the lower receiver are separated, we can turn our attention to the FCG. While not as simple as the MP40, its not very difficult to work with.

The trigger assembly is held in by the rear pin. The sear held in by the front pin.

The disconnector though, is a little pit of a pain to get out. It is held in by a small pin on the end of a springloaded plunger that must be depressed from the inside of the FCG housing to get to. I've found the easiest way to do this is to put a small punch in a vice, facing up. Put your plunger on the punch and push down. When the pin is exposed, take another small punch and lightly tap it out with a hammer.

I say lightly because it is a small pin, very tiny in fact, and if you aren't careful you can launch it into space. At that point, you are looking for a replacement pin.

Once the pin is out you simply let pressure off the FCG housing and the disconnector assembly should come right out. Your selector switch will be removable as well.

The parts of the FCG laid out in their respective areas. Not visible is the pin that holds in the disconnector housing.

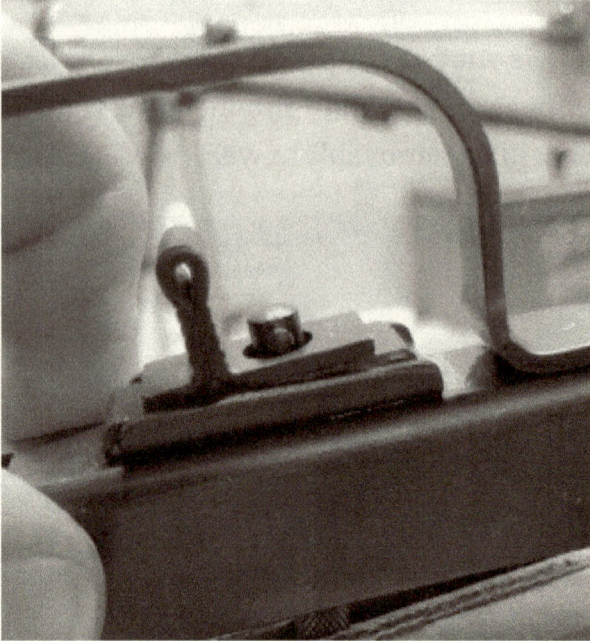

Here you can see the small pin that needs to be removed to take the disconnector assembly out. In this picture it is depressed from the inside of the FCG.

On the upper portion of the lower receiver you will find the ejector. It slides into the rear part of the magwell and is riveted in place just behind the magwell. A simple thing to replace, but of note, is that the rivet tail in the receiver area where the bolt rides must be flush with the ejector flat, other wise you will have "the issues" as the saying goes.

The small rivet head behind the magwell is the ejector rivet. Also shown here is the pin that holds in the magazine latch in the lower rear portion of the magwell.

The Bolt Group

The bolt group is just as easy to deal with as the rest of the gun consisting of the bold assembly, the recoil assembly and the buffer.

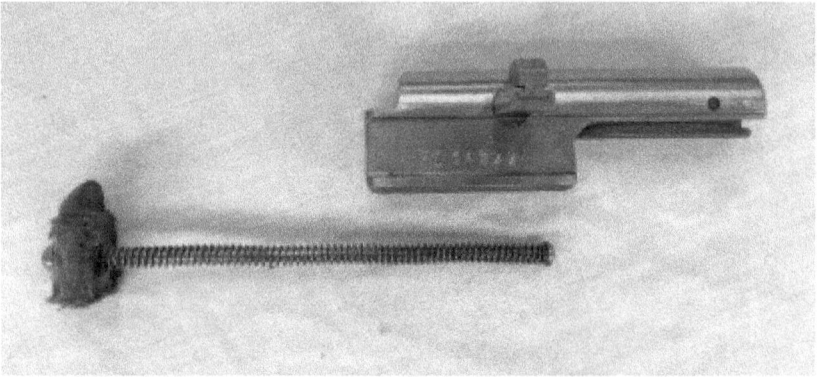

Here is the bolt group, not that the buffer is still on the recoil assembly.
Also note it is chewed to hell. It's time to replace my buffer.

Since I've just found out my buffer needs replacing, we will talk a bit about that.

The buffer in these guns take a lot of abuse. You have to figure it's getting hit by the back of the bolt at a thousand rounds a minute and the gun fires the 7.62x25mm round (unless the 9mm adapter is installed), which is a zippy little cartridge. They can get chewed up rather quickly. On a good note, they can take a ton of rounds before needing to be replaced. It depends on a couple things like how good your recoil spring is, smoothness of the gun, and even manufacturer of ammo. Basically, if you think it should be changed, change it out. They are plentiful and inexpensive. They even have new manufactured ones!

As to the recoil spring, it is good for several thousand rounds but if you think you need a new one go ahead and replace it. Again, they aren't very hard to find and are not expensive.

The bolt contains the extractor, extractor spring, firing pin and retainer pin. It also contains a safety similar to the MP40 in that it slides in and out in notches in the receiver. There is a notch for when the bolt is held in the open position, but I am here to tell you, don't trust it!

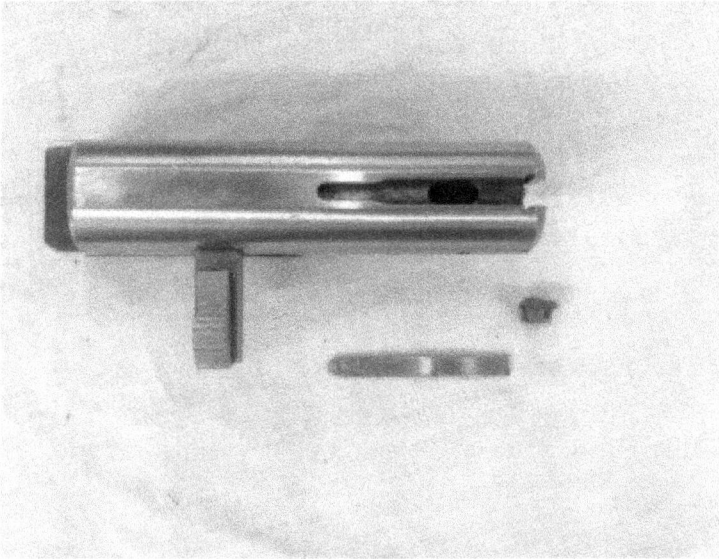

The bolt with the extractor and its holding spring removed.

The method for removing the extractor spring is to hold up the front portion over the extractor "ears" and push the spring forward using a punch in the rear hole of the spring. Underneath the spring you will see a slotted cutout. In that is the firing pin. That is removed by removing the pin located on the forward side of the bolt and reaching into the slot and driving the firing pin out the front of the bolt.

PPSH 41 Malfunctions and Repairs

It was touched on in my first book, but I will mention it again here. One of the problems that has come up is varying thickness of receiver sections in post sample guns that have been welded together. This is because the donor sections used vary in thickness, but not by much. You wont really notice it if you just take it for granted they are the same without measuring. Also, even if the correct thickness sections are used, they are welded in. Sometimes folks get lazy on their weld dressing and you can have a "hump" where it was welded.

Either way it amounts to the same thing, it slows the bolt down and you get light strikes.

The only thing you can do is take away metal to smooth things up.

You also may see a donor section that was too thin. In this case you may need to add metal to get things to line up correctly.

Other than that, you will more than likely just run into the run of the mill broken parts. Just swap them out and have fun

PPSH 41 Reassembly

Like the MP40 there aren't really any problem areas but there are some things to be aware of. When reinstalling the barrel, I have seen one that was notched twice for the install pin. Not thinking it made a difference, I installed the barrel and guess what? Yep, bullet strike in the front of the barrel jacket. Luckily it was on my post sample and I was able to fix it, but it did prove to be a valuable learning experience. Also, when installing the barrel, make sure you drive it in with a brass punch or a brass hammer hitting on the lower part of the barrel shoulder.

Firing the PPSH 41

A simple open bolt blowback subgun, it can be a handful when fired due to its high rate of fire.

For novice shooters this is especially true.

The gun tends to rise pretty fast when fired. If they are ready for it, it's not terribly bad.

Also don't forget the charging handle mounted safety. I've seen more then a few guys charge the bolt and accidentally engage it. Of course this has the action of either slowing the bolt down, so you get a light strike, or of having the bolt not move at all.

Either way, the look of bewilderment on the new shooters faces is a bit humorous.

One other quirky thing that you will see is that firing, extraction, and ejection can all happen perfectly and the gun will still jam. That's because the ejected brass goes straight up. Then comes straight down, right into the ejection port.

What a strange little gun right?

3 THE M3A1

"GREASE GUN"

The Thompson was used extensively during WWII. Chambered in 45 caliber; able to fire in either semi or automatic mode, and using a detachable magazine; it was a very good weapon, if a bit dated. Being a heavy weapon and extremely expensive to produce, the US military needed a replacement. Other contemporary sub machineguns of the time used few machined parts being comprised mostly of stamped metal.

The M3A1 "Grease Gun" was originally supposed to be a replacement for the Thompson. Making extensive use of stamped metal, it was both easier to produce and much lighter. It's been rumored that an M3A1 only cost 8 dollars to produce!

Today, a transferable M3A1 will set you back in the neighborhood of $20,000.00.

Thanks a lot Congress.

Having said all that, there are quite a few of these guns out there. Though they are crude and aiming can be a challenge, they are very sought-after weapons. So we may as well learn how to work on them.

M3A1 Basic Disassembly

1. Ensure the weapon is unloaded and the bolt is in the forward position.

2. While holding down the barrel retention spring turn barrel counter clockwise when viewed from the front.

3. With ejection port door open remove bolt and recoil group.

Once again, we see a simple field stripping operation. Why, oh why, can't we leave well enough alone?

It's because we have tools and it can come apart farther!

M3A1 Expanded Disassembly

Taking the "Greaser" apart to its smallest components really poses no challenges.

The painful part is putting it back together. Specifically, the FCG back in the receiver. Lets jump in the shallow end, shall we?

Bolt and Recoil Section

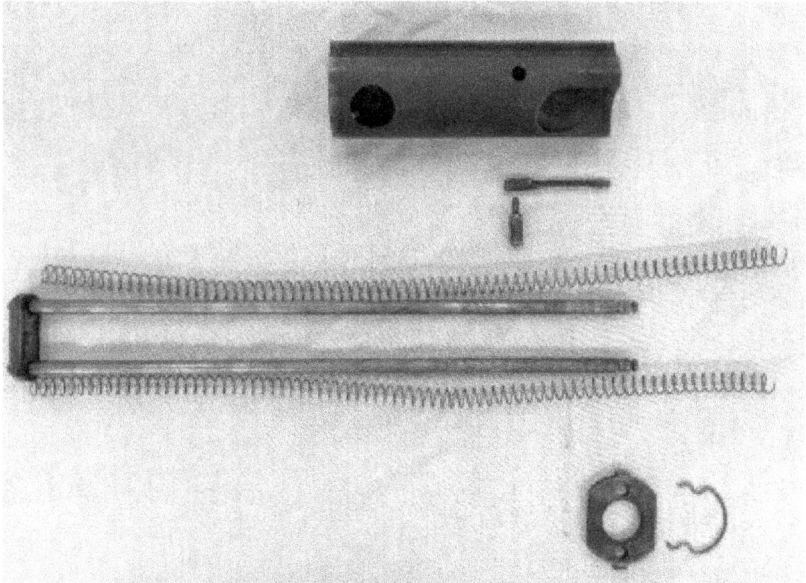

The components of the bolt and recoil section.

To disassemble the parts of the bolt and recoil section, start by pushing down on the bolt slightly with the back of the recoil spring guides on a flat surface.

Remove the spring metal retaining clip on top of the flat retaining ring, then remove the retaining ring.

Now while one hand holds the spring in their compressed spots, take off the bolt. Once that is done ease the springs up until they are no longer under compression.

To remove the extractor from the bolt simply drive out the pin from the bottom on the bolt. At that point push, the extractor out the front of the bolt using a long punch inserted into the small hole in the back of the bolt.

You have just completely taken apart the bolt and recoil group.

Receiver Section

The M3A1 receiver assembly taken apart into it's various components. Of note the buttstock is not shown.

To begin disassembly of the receiver assembly first remove the trigger guard. This is accomplished by pulling down and out on the rear of the trigger guard.

Once this is removed, take the ejector housing off by pulling down on the rear of it. At this point the magazine release can be removed by slightly moving it rearwards and pushing to the left of the receiver.

Next the sear pin is removed. That is the large pin in the front of the FCG housing.

Now the trigger pin can be removed. This is the double pin. The rear of it holds in the trigger. The front of it is a hold down for the sear.

Remove the FCG from the ejection port of the receiver. You will probably have to hold the receiver upside down and shake it a bit to get the FCG to fall into the correct position to remove it.

The Buttstock catch can be removed by driving the pin out and removing the parts. It of course, is spring loaded.

The ejection port cover can be removed, but you will have to take out the crimp placed there to keep the pin in. Be very careful if this is attempted, because if you try to pry the crimp out and the tool slips it may accidently hit the receiver. As it takes no small amount of effort to take this crimp out, that force will be transferred to the receiver and it may dent. That opens a whole new can of worms.

Short of a broken door there really is no reason to remove it so its generally best kept in place.

As to the ejector, it is riveted to the housing that was in front of the trigger guard. Replacement of this is just the two rivets but it would probably be just as easy to find a new housing ready to go.

This completes stripping the receiver.

M3A1 Malfunctions and Repairs

The M3A1 is a well-built little gun but it can have some issues. The first of which I haven't seen but is a bit troubling. That is the tendency for the rear of the receiver to split. Again, I haven't seen this, but I have heard of it happening from a few different people. Also, whether these were factory or post sample rewelds I have no idea. Its just something to look out for. Myself, I wouldn't be too alarmed over it if you are looking to purchase one.

The next thing I have seen is the tendency for the recoil spring guides to come apart. They are crimped in at their base and aren't supposed to come apart, but they don't know that and often do with enough use. I've seen them come apart but haven't seen that they caused any damage to the gun. Beyond a surprise when you take them apart. I suppose it could damage the rear of the receiver so that should be enough to fix em.

Obviously, replacement is the best option but if one can't be found I've had success with welding them in place. I weld from the outside and then clean it up. You just have to make sure that the welds don't stick out and they are flush, otherwise this causes the recoil spring guide to be at an angle or press on one area more than another on the base of the receiver.

Another thing to be aware of is the front barrel lock spring. This can get weak or bent and it's what prevents the barrel from rotating. Make sure that is good to go.

M3A1 Reassembly

As with the others there is no issues with saying reassembly is the reverse of assembly; save this:

To assemble the FCG in the gun it takes a little finessing.

First install the FCG into the gun from the ejection port. Then tilt the gun back until the FCG is almost in the right position. Push down on the sear with a finger till you can install the larger sear pin. Keep pressure on the sear, and with a punch, push down on the trigger until it goes in its slot. Again keeping pressure on the sear, line up the trigger pin hole. Push in the trigger pin as far as possible while pushing down on the sear. The other part of the trigger pin needs to hold down the sear so the sear must be kept in the down position when its installed.

Easy right?

Firing the M3A1

There are a couple quirks to this gun that I should mention. First the sights are horrible. But since they amount to little tabs that were just welded in place, I guess they are good enough.

Next comes the fun of charging the gun. If you load your magazine in first, you must be careful when charging the weapon. It has no charging handle, only a slot milled in the bolt. The idea is to place a finger or two, in the slot and pull back the bolt. Its easy enough to do but if your fingers slip or you don't pull back far enough to engage the sear, bang!

Now you finally get to fire the gun. Its iconic,. It's got a real cool slow rate of fire and it bounces as if you were firing the thing on horseback. Thats the best way I can describe firing the M3 Grease Gun..... Oh and the sights suck!

4 THE STEN MKII
"THE UGLY STICK THAT COULD"

I like to kid around when it comes to the Sten; to say things like "Its not pleasant to shoot", "Its ugly", or "If your mag is fully loaded, the gun turns sideways", etc.

Well I'm here to tell you it's all of that and more!

Seriously though this little gun was simple, cheap to build, could take a hell of a lot of abuse and still work. I've only ever seen one break and that was due to a broken trigger spring.

The brits literally made millions of Stens and you can still find them in use today in some third world countries.

Take that, AK47!

Sten MKII Basic Disassembly

Ensure the gun is unloaded and bolt is in the forward position.

1. Push button (rear spring cap) and slide buttstock down off the receiver.

2. Push in rear spring cap till it stops, rotate the spring cap counterclockwise. While still keeping pressure on spring cap, let it rise until spring is no longer compressed. Remove the cap and spring.

3. Pull back bolt until charging handle can be removed in the "safe" area of the charging handle slot by lifting out.

4. While depressing trigger, remove the bolt from rear of receiver.

5. Pull back magazine housing plunger and rotate barrel jacket counter clock wise, when viewed from the front until it can be removed from gun. Remove barrel jacket and barrel.

Again, not much to it. In fact, the last step doesn't have to be done for normal cleaning. It just makes it easier.

Once again, since we can't leave well enough alone, we are going to take it down even further.

Sten MKII Expanded Disassembly

The Sten MKII, being select fire, has a few more do dads (that's a technical term by the way) than some others but it is still easy to strip. There are two things that are held in place by cotter pins, the selector switch and the lock that holds the magwell from rotating. If you don't have to remove these, don't, especially the magwell rotation lock. Aside from that its easy. So, lets grab our hammer and see what we can do!

Bolt Section

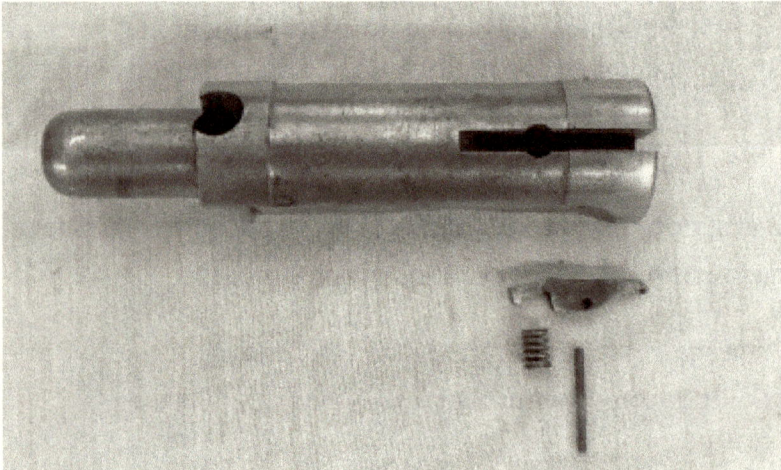

That's it, the whole bolt assembly.

It gets pretty complex here so hold on. Take a punch and drive out the extractor pin. Remove the extractor and spring.

You have just disassembled the Sten MKII bolt.

One thing of note though, in the extractor slot you can expect to find a ton of gunk, dirt, carbon, small trees, etc. Make sure you clean that out.

Receiver Section

The Sten MKII receiver broken down. The only items not removed are the magwell and magwell lock. On this gun the front sight is welded on and would need to be removed in order to remove the magwell. Also the magwell is in the lower stowed position, not in the firing position.

To strip the receiver section, remove the bolt to the FCG cover. Remove the cover and lay aside.

Unhook the front of the trigger return spring from the hooked end of the tripping lever pawl. That's the fatter spring on the sear. Once this has been done, remove the tripping lever pawl itself by sliding off its mounting bar on the sear.

Remove the trigger pin then remove the sear pin while holding onto the bar on the sear.

Remove the trigger, tripping lever, and sear at the same time.

The selector switch is held in by a cotter pin and normally does not need to be removed. If you do need to remove the selector remove the cotter pin and take the selector out through the "A" side (right hand side of gun when viewed from rear).

To remove the magazine release, first take out the slotted screw that holds on the magazine catch retainer. Then slide the retainer up and off. The magazine catch can now be removed by twisting it off towards the top of the receiver. When doing this, be prepared, the magazine catch spring could fly off to the great expanse.

To remove the magwell you must first remove the front sight. This next step is optional but makes it easier.

After removing the front sight, the cotter pin from the magazine housing plunger must be removed.

Be advised the plunger is under spring pressure and will fly up when the cotter pin is removed.

Once the cotter pin is removed take off the plunger, spring, and the barrel nut catch. You can now remove the magwell by sliding it off the front of the receiver.

You have now disassembled the MKII receiver.

Sten MKII Malfunctions and Repairs

Honestly, I haven't seen many of these come in for repairs and that's because they are build like a tank, the receiver tube is decently thick to resist issues and the rest is just as bullet proof. Of course, they can still be dented/deformed with enough force but everything is just built robustly in general.

The only things I've seen them come in for is broken springs in the FCG, a screwed-up extractor and mag issues. As with all broken stuff, remove and replace or "R-squared" as I like to say.

The mag issue is really annoying as they are overly sensitive to magazines. In total, I've worked on 6 of these guns and every one of them had a sensitivity to mags. Not just a "won't feed" issue but a "won't fit" issue. Why this, I have no idea. Maybe it's that there were a few different mag makers and they all wanted to be different or maybe it was just a series of Fridays, who knows.

Long story short, if your gun isn't working with one mag try another. Of course, if you buy them online be prepared for 3 of the 10 you ordered not to fit. It's kind of frustrating.

Sten MKII Reassembly

Assembling the Sten MKII is fairly easy, it's the reverse of disassembly, with only a couple items that need to be addressed.

The first of which is the magazine catch. To reinstall the magazine catch onto the magwell, place the hooked end of the magazine catch into its slot. With a flat object compress the spring. I use a punch but a "flat object" would be easier. Unless said flat object is a stamp. Then the process would be harder. Slide the magazine catch onto the magwell and install the magazine catch retainer. Pop in the screw and done.

The other area is the FCG. Installation, in order, is as follows. Install the Selector switch. With receiver placed on its back install the trigger and tripping lever. Put in the trigger pin. Pull up on the tripping lever and install the sear. The tripping lever goes between the bottom flat area of the sear and the bar where the tripping lever pawl seats. Place pin in sear once it has been positioned in the FCG area. Put on the tripping lever pawl with the hook facing forward. Install the trigger spring on the trigger hook then onto the hook on the tripping lever pawl.

Really the worst part is installing the trigger spring on the trigger hook but just get it underneath the hook and force it up with a small hooked awl.

Firing the Sten MKII

The first thing you will realize is that there is no comfortable way to hold a fully loaded Sten gun. Some grip it high up on the receiver. Some grip it by the magazine. Still others, myself included, grip it close in on the FCG housing. Either of these will work so do whatever feels "comfortable" to you.

Secondly, you will notice a fully loaded Sten gun tends to tilt to the left due to the weight of the fully loaded magazine. It's a weird feeling but you can get used to it in short order.

Once these hurdles are overcome you can finally fire the damn thing. And you find out where this gun truly shines; as a hammer. Indeed, it bounces around so much I think if you held the buttstock on a nail and pulled the trigger it really would drive the nail in quite effectively!

5 FINAL THOUGHTS

You have probably noticed that this is a short book. That's because there just isn't much to talk about on these guns. They are overly simple for a reason. Complex things, especially in a war, tend to break!

But break they do and that is why this book is here.

Most blow back guns follow the same formula as these four guns listed here. Want to work on a KP44? No problem, its basically a stamped PPSH. How about a Swedish K? No sweat, the FCG is a bit more complex than the ones in the book but not so much so that what we have here doesn't compare. The point is, with these four, you can see how all the other blow back guns operate. Remember these books are designed to get you to start thinking about how to fix all this stuff. Hell, I'd have to live forever to write a book on each particular gun you may see.

Anyway hope ya'll enjoyed the second book in the machine gun maintenance line. I promise the next one will be even better!!